by Maria Beuck
illustrated by Courtney Hopkinson

SCHOOL PUBLISHERS

Copyright © by Harcourt, Inc.

All rights reserved. No part of this publication may be reproduced or transmitted in any form or by any means, electronic or mechanical, including photocopy, recording, or any information storage and retrieval system, without permission in writing from the publisher.

Requests for permission to make copies of any part of the work should be addressed to School Permissions and Copyrights, Harcourt, Inc., 6277 Sea Harbor Drive, Orlando, Florida 32887–6777. Fax: 407-345-2418.

HARCOURT and the Harcourt Logo are trademarks of Harcourt, Inc., registered in the United States of America and/or other jurisdictions.

Printed in China

ISBN 10: 0-15-350439-0
ISBN 13: 978-0-15-350439-6

Ordering Options
ISBN 10: 0-15-350332-7 (Grade 2 Below-Level Collection)
ISBN 13: 978-0-15-350332-0 (Grade 2 Below-Level Collection)
ISBN 10: 0-15-357447-X (package of 5)
ISBN 13: 978-0-15-357447-4 (package of 5)

If you have received these materials as examination copies free of charge, Harcourt School Publishers retains title to the materials and they may not be resold. Resale of examination copies is strictly prohibited and is illegal.

Possession of this publication in print format does not entitle users to convert this publication, or any portion of it, into electronic format.

4 5 6 7 8 9 10   985   15 14 13 12 11 10 09 08

Characters

Announcer

Host

Mystery Guest

Player 1

Player 2

Setting: The set of a television game show

Announcer: Welcome to the game show *What's My Pet?*. Here to entertain you is our carefree game show host, Cliff Dear.

Host: Hello, everyone. Our sleuths are eager to get started!

Announcer: Here are the rules. Players take turns asking the Mystery Guest questions about his pet. After he responds, the players guess what the pet is.

Host: You have two guesses each. The player who guesses accurately wins. Are you ready to play?

Players 1 and 2: We're ready!

Host: Then let's welcome our Mystery Guest. Welcome!

Mystery Guest: Thank you, Cliff.

Host: The first question, please.

Player 1: Mystery Guest, is your pet kept in a cage?

Mystery Guest: Yes, it is.

Player 2: I know what it is!

Host: You should ask a question first, Player 2!

Player 2: Mystery Guest, does your pet have feathers?

Mystery Guest: No, it doesn't.

Player 2: Oh! Then I'm wrong. I don't know what it is.

Announcer: Our committee of sleuths will have to concentrate. Who will earn the prize?

Host: Back to you, Player 1.

Player 1: The pet is kept in a cage. It doesn't have feathers. Let me think.

Host: Try to be creative!

Announcer: Be brave! Take a risk!

Player 1: Mystery Guest, is the cage kept in your house?

Mystery Guest: Yes, it is.

Player 2: The only pet I know that is kept in a cage inside a house is a . . .

Player 1: MOUSE!

Mystery Guest: Yes! My pet is a mouse.

Player 2: I was going to say that. I'm the winner.

Announcer: Sorry, Player 2, we can only have one winner.

Host: Player 1, you are our super sleuth. Your prize is a statue of a mouse!

Player 1: I'm thrilled!

Host: This is Cliff Dear, saying good-bye for another week. Join us again next time!

Think Critically

1. What was the setting of the Readers' Theater?

2. In the book, on what page do you find out what the prize was?

3. What was the Mystery Guest's pet?

4. What pet do you think Player 2 was going to guess when he changed his mind?

5. Do you think *What's My Pet?* was a good game show? Why or why not?

 **Social Studies**

Write a Paragraph The Readers' Theater is about a television game show. Write a paragraph about ways people were entertained before television was invented.

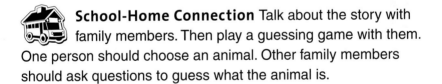 **School-Home Connection** Talk about the story with family members. Then play a guessing game with them. One person should choose an animal. Other family members should ask questions to guess what the animal is.

Word Count: 324